Radio Frequency Identification Technologies

A Workshop Summary

Committee on Radio Frequency Identification Technologies

Computer Science and Telecommunications Board

Division on Engineering and Physical Sciences

NATIONAL RESEARCH COUNCIL
OF THE NATIONAL ACADEMIES

THE NATIONAL ACADEMIES PRESS
Washington, D.C.
www.nap.edu

THE NATIONAL ACADEMIES PRESS 500 Fifth Street, N.W. Washington, DC 20001

NOTICE: The project that is the subject of this report was approved by the Governing Board of the National Research Council, whose members are drawn from the councils of the National Academy of Sciences, the National Academy of Engineering, and the Institute of Medicine. The members of the committee responsible for the report were chosen for their special competences and with regard for appropriate balance.

Support for this project was provided by the Defense Advanced Research Projects Agency (DARPA). Any opinions, findings, conclusions, or recommendations expressed in this publication are those of the authors and do not necessarily reflect the views of the agency that provided support for the project.

International Standard Book Number 0-309-09543-3 (Book)
International Standard Book Number 0-309-54778-4 (PDF)

Cover designed by Jennifer M. Bishop.

Additional copies of this report are available from:

The National Academies Press
500 Fifth Street, N.W., Lockbox 285
Washington, DC 20055
(800) 624-6242
(202) 334-3313 (in the Washington metropolitan area)
http://www.nap.edu

Printed in the United States of America

THE NATIONAL ACADEMIES

Advisers to the Nation on Science, Engineering, and Medicine

The **National Academy of Sciences** is a private, nonprofit, self-perpetuating society of distinguished scholars engaged in scientific and engineering research, dedicated to the furtherance of science and technology and to their use for the general welfare. Upon the authority of the charter granted to it by the Congress in 1863, the Academy has a mandate that requires it to advise the federal government on scientific and technical matters. Dr. Bruce M. Alberts is president of the National Academy of Sciences.

The **National Academy of Engineering** was established in 1964, under the charter of the National Academy of Sciences, as a parallel organization of outstanding engineers. It is autonomous in its administration and in the selection of its members, sharing with the National Academy of Sciences the responsibility for advising the federal government. The National Academy of Engineering also sponsors engineering programs aimed at meeting national needs, encourages education and research, and recognizes the superior achievements of engineers. Dr. Wm. A. Wulf is president of the National Academy of Engineering.

The **Institute of Medicine** was established in 1970 by the National Academy of Sciences to secure the services of eminent members of appropriate professions in the examination of policy matters pertaining to the health of the public. The Institute acts under the responsibility given to the National Academy of Sciences by its congressional charter to be an adviser to the federal government and, upon its own initiative, to identify issues of medical care, research, and education. Dr. Harvey V. Fineberg is president of the Institute of Medicine.

The **National Research Council** was organized by the National Academy of Sciences in 1916 to associate the broad community of science and technology with the Academy's purposes of furthering knowledge and advising the federal government. Functioning in accordance with general policies determined by the Academy, the Council has become the principal operating agency of both the National Academy of Sciences and the National Academy of Engineering in providing services to the government, the public, and the scientific and engineering communities. The Council is administered jointly by both Academies and the Institute of Medicine. Dr. Bruce M. Alberts and Dr. Wm. A. Wulf are chair and vice chair, respectively, of the National Research Council.

www.national-academies.org

COMMITTEE ON RADIO FREQUENCY IDENTIFICATION TECHNOLOGIES

GAETANO BORRIELLO, University of Washington, *Chair*
DANA CUFF, University of California, Los Angeles
CHRIS DIORIO, University of Washington
BILL SCHILIT, Intel Corporation
STEVEN SHAFER, Microsoft Corporation
PAUL ZIPKIN, Duke University

Staff

LYNETTE I. MILLETT, Senior Program Officer
DAVID PADGHAM, Research Associate (until May 7, 2004)
PHIL HILLIARD, Research Associate (until June 4, 2004)
JANICE SABUDA, Senior Program Assistant

COMPUTER SCIENCE AND TELECOMMUNICATIONS BOARD

For more information on CSTB, see its Web site at <http://www.cstb.org>, write to CSTB, National Research Council, 500 Fifth Street, N.W., Washington, DC 20001, call at (202) 334-2605, or e-mail CSTB at cstb@nas.edu.

Preface

The day when each discrete manufactured object in our everyday environment comes with an embedded computer chip is arguably getting closer. Radio frequency identification (RFID) technology is currently one of the most powerful forces moving us in that direction. RFID technology uses three main components: (1) a microchip with a radio antenna (often referred to as an RFID tag), (2) a device to send and receive a signal from such tags (called an RFID tag reader), and (3) the hardware and software environment that enables useful information to be derived from the interactions of tags and readers.

The technology has already shown promise in uses involving transportation (for example, SmarTrip and E-ZPass for parking fees or transit fares and highway tolls) and commerce (for example, Mobil's SpeedPass). A number of other uses for the technology are under development and in some cases deployed, including applications such as real-time inventory management and "smart" checkout in stores and libraries. Current technical issues with the technology include such matters as the size and production cost of individual tags, interference between readers and other devices in their spectrum range, and the effective range of tags and readers.

Many industry leaders look forward to the benefits and cost savings that RFID technology might bring to their operations. However, on the consumer and regulatory side, there are many concerns and unanswered questions about the technology—for example, what are the ramifications for personal privacy of embedding RFID tags in consumer products? Indeed, more than one company has had to change or rethink its plans for RFID technology because of the concerns of consumers and privacy advocates about how the technology would be used.

Currently, RFID technology seems to be at a crucial point—in the development of the technology itself, on the one hand, and in the development of the policies and standards that will affect its use and deployment, on the other. In addition, with the recent entrance into the RFID arena of two major participants—the U.S. Department of Defense and the nation's largest retailer, Wal-Mart—the technology may be on its way to becoming ubiquitous in American society.

As a follow-on activity to the project that produced the report *Embedded Everywhere: A Research Agenda for Networked Systems of Embedded Computers*,[1] the Computer Science and Telecommunications Board (CSTB) of the National Research Council (NRC) conducted a short workshop that explored RFID technology and related technical and policy issues. Workshop participants included representatives from industry, academia, government, and relevant non-governmental organizations.

---

[1] See National Research Council, 2001, *Embedded, Everywhere: A Research Agenda for Networked Systems of Embedded Computers*, National Academy Press, Washington, D.C.

To conduct the workshop, the NRC appointed a steering committee—the Committee on Radio Frequency Identification Technologies—with expertise in the following areas: technical and engineering aspects of RFID and related technologies; the practical and business uses of these technologies; the implications of RFIDs for personal privacy, anonymity, and so on; and the policies, standards, and regulations surrounding RFIDs. The committee developed the workshop agenda, participated in the workshop, and composed this workshop summary report. The committee met in person twice (during and immediately after the workshop), as well as via teleconference. Committee members also did a great deal of work electronically via e-mail.

This report is the committee's synthesis of key points made in presentations by the workshop panelists and in subsequent discussions. Although the summary is a report prepared on the basis of presentations and discussions at the workshop and among committee members, the comments do not necessarily reflect the views of the committee, nor are they findings or recommendations of the National Research Council.

The committee thanks the individuals who contributed to its work, including the workshop panelists (listed in the workshop agenda in Appendix A) and participants. The committee appreciates their willingness to address the questions posed to them and is grateful for their insights. The reviewers of the draft report provided insightful and constructive comments that contributed significantly to the clarity of the report.

Neither this report nor the workshop itself would have been possible without the dedicated and professional efforts of CSTB staff. David Padgham was involved in organizing the event and was instrumental in bringing the excellent collection of panelists together. The logistics of the event were flawless, thanks to Janice Sabuda. Phil Hilliard provided excellent notes on the workshop discussions. Dorothy Sawicki from the Division on Engineering and Physical Sciences' editorial staff made significant editorial contributions to the final manuscript. Extra special thanks go to Lynette Millett, who developed the idea for this workshop and went beyond the call of duty in keeping overcommitted members of the steering committee on task. Without her, none of this would have been possible.

Gaetano Borriello, *Chair*
Committee on Radio Frequency Identification Technologies

Acknowledgment of Reviewers

This report has been reviewed in draft form by individuals chosen for their diverse perspectives and technical expertise, in accordance with procedures approved by the National Research Council's (NRC's) Report Review Committee. The purpose of this independent review is to provide candid and critical comments that will assist the institution in making its published report as sound as possible and to ensure that the report meets institutional standards for objectivity, evidence, and responsiveness to the study charge. The review comments and draft manuscript remain confidential to protect the integrity of the deliberative process. We wish to thank the following individuals for their review of this report:

Bruce Eckfeldt, Cyrus Innovation
Deborah Estrin, University of California, Los Angeles
Randy H. Katz, University of California, Berkeley
Ravi Pappu, ThingMagic
Gregory J. Pottie, University of California, Los Angeles
Sumit Roy, University of Washington
Fred B. Schneider, Cornell University
William Stead, Vanderbilt University

Although the reviewers listed above have provided many constructive comments and suggestions, they were not asked to endorse the conclusions or recommendations, nor did they see the final draft of the report before its release. The review of this report was overseen by Richard Rowberg, Division on Engineering and Physical Sciences, National Research Council. Appointed by the National Research Council, he was responsible for making certain that an independent examination of this report was carried out in accordance with institutional procedures and that all review comments were carefully considered. Responsibility for the final content of this report rests entirely with the authoring committee and the institution.

Contents

Overview

This report summarizes the main points made in the presentations and subsequent discussions at a workshop on radio frequency identification (RFID) technologies held May 10-11, 2004, in Seattle, Washington, under the auspices of the Committee on Radio Frequency Identification Technologies of the National Research Council's Computer Science and Telecommunications Board.

Radio frequency identification is a generic term for a set of technologies that use radio frequency (RF) to communicate data (a central component of which is an identity—specifically, a unique number). In its most rudimentary form, an RFID system consists of a *tag,* which includes the identity, attached to an object and a *reader* that can query the tag to find out what that identity is. The technology has been advancing over the past several years, and the application space has been broadening. RFID has been used for a range of activities from pinpointing the position of runners in marathons to tracking livestock to automating supply chains and assisting in inventory management for major retail vendors.

The workshop agenda and panel topics were developed by the steering committee to provide for a broad discussion covering not only the technical aspects of the technology but also its applications and, as importantly, its implications for society. Although a short workshop cannot do justice to the complexity of all these aspects of RFID technologies, it does serve as a good starting point for gaining an understanding of the basics of some of the major issues. Listed below are the topics addressed by panels at the workshop (see Appendix A for the names of the panelists):

- Brief History and Overview of RFID Technology—Where We Stand
- Business Case for and Against RFID Technologies
- Where the Technology Is Going
- RFID Infrastructure and Data Management Issues
- Privacy, Social, and Cultural Concerns
- RFID, Government, and Standards
- Looking to the Future—Predictive and Speculative

Following are a few of the main themes arising from the workshop discussions and panel sessions. They do not constitute conclusions or findings of the committee; instead these themes incorporate ideas extracted from the workshop that came through strongly during discussions.

- RFID technologies come in many variations and exhibit a range of capabilities; understanding the specifics of a particular RFID technology is important to determining appropriate uses and applications for it.
- In many ways, RFID is still in its infancy. Much experimentation and study must be done in order to achieve a deep understanding of its potential and implications.
- The cultural and social questions that arise from the use and deployment of RFID technologies include significant challenges in the areas of privacy and data collection.

A summary of the panel discussions, together with background research and insights from the steering committee, is presented in this report. Chapter 1, "Technology and Applications," addresses technical constraints, architecture, standards, and business applications. Chapter 2, "Society and Culture," discusses some of the potential social and policy implications of RFID.

1

Technology and Applications

Identification is a powerful capability, useful in classifying, counting, and organizing objects. These operations are essential to many aspects of modern life, including manufacturing, the logistics of distribution, and the various stages of supply chains, and they operate on scales ranging from the level of the individual consumer to that of global trade. In the past, identification was done visually—by observing characteristics of objects. When copies of manufactured objects that are essentially identical have to be identified, distinguishing markings have been added. Efficient and accurate means are needed to recognize the markings and thus determine the identity of the marked objects. Therefore, an identification system consists of identifying markings and readers of those markings. The first readers were human beings; technical innovations subsequently resulted in photodetectors, cameras, and lasers being used as readers. The markings have evolved into the popular bar code that is printed on almost every package and item.

Radio frequency identification (RFID) is a means of identifying objects by interrogating a unique characteristic of the object (such as a unique identifying number stored on a silicon chip attached to the object) using radio waves.[1] This technology promises orders-of-magnitude greater efficiency and accuracy than were possible previous technologies. Although RFID is not a recent development,[2] advances in semiconductor technology have now made this method practical and much more cost-effective. RFID has many advantages over visual markings—primary among them the ability to identify objects without the requirement of line of sight. This means that objects can be identified even when they are tightly packed together or their surface markings are removed, marred, or obscured.

The elements added to objects to facilitate identification in this way are called RFID tags. Tags consist of at least two basic subsystems: (1) a memory element that holds an identification

---

[1] For more background information on RFID, see Roy Want, 2004, "RFID: A Key to Automating Everything," *Scientific American*, January, pp. 56–65. See also Brian Dipert, 2004, "Reading Between the Lines: RFIDs Confront the Venerable Bar Code," *EDN Magazine,* October 14, available online at <http://www.edn.com/index.asp?layout=articlePrint&articleID=CA468418>, accessed December 14, 2004.

[2] Indeed, one workshop participant argued that RFID has been around since 1886, with Hertz's experiments in radio frequency propagation over a 1 meter range and developed into its first practical application in a half-ton tag used to identify friend or foe in aircraft in 1942. For comparison, bar codes were invented at Drexel University in 1948 and started becoming practical in 1962 with the advent of laser readers for codes printed with inexpensive ink. See also Harry Stockman, 1948, "Communication by Means of Reflected Power," in *Proceedings of the IRE* [Institute of Radio Engineers], October, pp. 1196-1204, for an early description of the theory and implementation of RFID.

number (a string of binary values) or some other identifying characteristic, and (2) an antenna to radiate or reradiate radio frequency (RF) energy, modulated by the identification number, to an apparatus that can detect that modulation and thus the identification value. Many variations of these two elements are possible, giving RFID tags quite a wide range of capabilities, as discussed below.

RFID VARIABLES

This section provides an overview of some of the components of an RFID system. It is important to note that, although many variations are available in each of the elements of an RFID tag, the variations are not necessarily available in all combinations. For example, a tag that can be read from a long way off will most likely require its own power source; a tag with no battery may be limited to a range of a few tens of meters. Some of the parameters to consider when evaluating or analyzing RFID systems are power requirements, the method of coupling between readers and tags, the receiving sensitivity and power output of antennas, the power requirements of the RFID tag chip (if the identifying tag uses a chip), and the frequency of operation. Several choices are usually available for each of these parameters, but the field of available tags is not simply the outer-product of all these options, because some combinations are not technically feasible or cost-effective.

Tags

The Basic RFID Tag

The simplest version of an RFID tag is a passive identification (ID) tag. It does not contain its own power source but instead harvests the power it needs from the reader's RF emissions.[3] It holds only a unique identifier and no other state information. When a reader reads the tag's ID, it typically uses the ID to index a database that contains more expansive information about the object. For example, a tag on a package of pharmaceuticals may point to a database entry about the provenance of the drugs in the package, the distribution history of the package, and its final destination. As another example, electronic-article surveillance (EAS) systems currently employed extensively in libraries use the physical characteristics of a magnetic ribbon to backscatter a unique signature.

The coupled design of the reader and the tag antenna determines the range at which the tag's ID can be read. Since this basic RFID tag is entirely dependent on getting power from the reader, the range with today's technology tends to be quite limited, varying from near contact (so-called contactless technologies—such as some smart cards) to a maximum of about 15 meters (for many of the tags currently in supply-chain trials). Other complications in the reading process include the presence of multiple tags, interference from other radio sources (and in particular other readers), the absorption of radio energy by different materials between the tag and the reader,[4] and the fact that reader transmission power is limited by regulatory bodies (see below).

---

[3] A reader communicates information to a tag by modulating an RF waveform, typically using amplitude-shift-keying (ASK) modulation. A reader receives information from a tag by transmitting a continuous-wave (CW) RF signal to the tag; the tag responds by modulating the radar cross section (the impedance match) of its antenna, thereby backscattering an information signal to the reader.

[4] RFID systems use either reader-talks-first or tag-talks-first operation. In reader-talks-first operation, tags wait to receive commands from a reader before backscattering. A tag responds with an information signal by modulating its antenna impedance only after being directed to do so by a reader. In tag-talks-first operation, tags backscatter information to a reader as soon as the tag enters an energizing RF field. In this latter case, a tag modulates its antenna impedance with an information signal until being directed to stop doing so by a reader. In addition, RFID systems can be half-duplex or full-duplex, with the former being

For certain frequencies of operation, particularly problematic barriers between tags and readers are metal and water (present in large quantities in human bodies), making it difficult to tag many individual items found in supermarkets.

Tags with Extended Memory

Expanding the memory capacity of an RFID tag allows the object to store data about itself in addition to its ID. Extended memory is a particularly useful capability when a tag is read by a reader that is not connected to a database of information about that tag. This disconnection might be due either to limited network connectivity or limited access (for example, as a result of being in a different administrative domain without access rights to that database).

Tags with Sensing Capability

Adding even a simple sensor to an RFID tag can radically increase its utility. Even if only a single bit of the tag's data comes from an integrated sensor, it can radically change usage models. For example, a sensor to detect whether a package has been opened can be as simple as a thin wire that gets cut when the package is opened, thereby toggling a bit in the tag's identifying number. A temperature-threshold sensor can inform a reader that a tag at some point reached a temperature higher than recommended—a useful capability in food and drug distribution.

Tags with Their Own Power Sources

Enhancing the communication range of RFID tags opens up many more applications. Enhancing range can be most easily accomplished by giving tags their own power sources. Of course, the capacity of the batteries used will have a large impact on the usage models for this class of tags. Depending on how often a tag is asked to transmit, its local battery may last anywhere from days to years. However, a self-contained power source allows the tag to have more interesting sensors that, in addition, can be used even when the tag is not near a reader to repeatedly sample some aspect of the environment at regular intervals.

Tags That Can Communicate with Other Tags

At the opposite end of the spectrum from basic tags that can only supply an ID to a reader are RFID tags that can communicate in a peer-to-peer fashion under their own power. These (along with some kinds of tag readers) can be thought of as nodes of a sensor network (the subject of a recent study by the National Research Council's Computer Science and Telecommunications Board[5]).

Tag Costs and Materials

Today's RFID tags vary in cost from a fraction of a U.S. dollar (read-only, passive tags) to several hundred U.S. dollars (active tags with their own power source and sensing capability). Tag costs are dominated by the interconnection of the silicon chip and antenna and their assembly into a package. They can weigh fractions of a gram (consisting of a small silicon chip and a thin antenna on mylar substrate). Ranges vary from 0.02 meter to 1 meter for near-field tags, to 15 meters for far-field tags, and up to 300 meters for high-end active tags, with every range in between also available. The antenna is the largest component of a tag. Tag antennas vary from less than a millimeter for the smallest passive tags (in this case the tag is on the silicon chip) to dozens of centimeters for some far-field tags. Most antennas are made of flexible material,

---

more common. With half-duplex, readers and tags do not talk simultaneously; rather, readers talk and tags listen, or vice versa.

[5] See National Research Council, 2001, *Embedded, Everywhere: A Research Agenda for Networked Systems of Embedded Computers*, National Academy Press, Washington, D.C., for a more complete discussion of sensor networks.

allowing some shaping to different form-factors. Other tags are packaged in epoxy or glass containers so that they can be subjected to washing or injected into animals.

Technologies Adjacent to Tags

Technologies adjacent to RFID tags include the following: sensor networks whose nodes are similar to active tags; contactless smart cards, which are similar to near-field tags but have a bit more processing power; and bar codes, the cheapest tagging technology (just ink on paper). Which sorts of tags or sensors to use is an important consideration for application developers. Variables that should be taken into account include these: read range, orientation independence, operating frequency, multitag reading (rather than one at a time), fast read rates (for example, tags that can function while traveling at high speeds, as when paying highway tolls automatically), resilience to ambient noise and interference, readability when tags are concentrated in a small space, reliability, and maintenance cost (which can vary over a wide range for all tag types, but especially for active tags). Not all of these characteristics are achievable to the desired degrees; some still fall within the realm of open research problems.

Although the difference between peer-to-peer tags and sensor network nodes is not sharply defined, certain characteristics of RFID tags usually distinguish them from sensor network nodes. RFID tags tend to communicate directly to readers (in a star pattern) and do not emphasize the routing of data through a network of peer nodes, as in ad hoc sensor networks. Usage models for tags do not rely on extensive computational capability within the tag itself; instead the reader is expected to provide a conduit to large-scale computational and storage resources. In sensor networks, nodes may be required to cache large amounts of collected data and elaborate them (often with complex digital signal processing algorithms) before communicating summary results to other nodes.

As indicated above, the term *RFID* is used to refer to a large spectrum of tag capabilities. Clearly, such a wide range leads to a need to standardize so that there is some hope for interoperability between large sets of tags and readers. As discussed in the section below entitled "Standards Bodies and Standards," early work toward standardization was begun by the International Organization for Standardization (ISO) and by the Massachusetts Institute of Technology's (MIT's) Auto-ID Center. The Auto-ID work is now being carried on by an industry consortium known as EPCglobal. EPCglobal has developed a taxonomy of tag classes (see Box 1.1), as well as standard RF signaling protocols between tags and readers, and formats for the storage of identity and data in tags.

Readers and Reader Infrastructure

Readers are the elements complementary to tags in RFID systems, as described in the subsections below. Readers must match the specific needs of tags and act as arbitrators when more than one tag is within their range. Readers communicate with tags and act as the tags' gateways to other information systems, such as databases that provide data indexed by the read identity. Readers must also coordinate among themselves at multiple levels, as they may interfere with one another's transmissions when they are in close proximity and need to share the spectrum. Also, they need to get the data that they read to the appropriate destinations so that the data are consistent when accessed by applications—which may involve transferring data from reader to reader as though the readers were nodes in a larger network.

BOX 1.1

Definitions of Classes of RFID Tags

Within 900 megahertz (MHz) radio frequency identification (RFID), the Auto-ID Center created the following class structure. Although this structure has not been formally adopted by the International Organization for Standardization (ISO), its use is continuing within EPCglobal, an industry consortium. A brief overview of this structure was presented at the workshop and is summarized below.

Class-1: Identity Tags (normative)

Class-1 identity tags are passive-backscatter read/write tags with the following minimum features:

- An electronic product code (EPC) identifier,
- A tag identifier (TID),
- A "kill" function that permanently disables the tag,
- Optional password-protected access control, and
- Optional user memory.

Class Restrictions (normative)

Class-2, Class-3, Class-4, or higher class tags shall not conflict with the operation of, nor degrade the performance of, Class-1 tags located in the same radio frequency environment.

Higher-Class Tags (informative)

The following class descriptions provide an example of how higher-class tag features might be delineated:

- **Class-2: Higher-Functionality Tags**—Passive tags with the following anticipated features above and beyond those of Class-1 tags:
  —An extended TID,
  —Extended user memory,
  —Authenticated access control, and
  —Additional features (TBD) as will be defined in the Class-2 specification.
- **Class-3: Semipassive Tags**—Semipassive tags with the following anticipated features above and beyond those of Class-2 tags:
  —An integral power source, and
  —Integrated sensing circuitry.
- **Class-4: Active Tags**—Active tags with the following anticipated features above and beyond those of Class-3 tags:
  —Tag-to-tag communications,
  —Active communications, and
  —Ad hoc networking capabilities.

Readers of Basic Tags

Reader antennas are designed to radiate energy to tags. For tags that contain an integrated circuit, the tag's power-harvesting elements must be able to collect enough energy to power the tag's chip and modulate the reflected signal. The reader must be sensitive enough to pick up this returned signal and interpret it. It is important to note that in passive tags the returned power falls off as the *fourth* power of the distance, that is, proportionally to $1/d^4$, where d is the distance from the reader to the tag. This is because the tag returns only some of the power that reaches it. Thus, the typical $1/d^2$ falloff of RF power is squared—meaning that to double the distance from reader to tag requires 16 times the power—because the RF transmissions must go round-trip.

Limiting factors on readers include the following: the amount of power that the reader can radiate (it is government-regulated), the reader's receive sensitivity (it is cost-sensitive), the reader's antenna gain (government-regulated), the size of the tag's antenna (there are cost and size considerations), the power requirements of the tag (they involve silicon processing), and constraints on the silicon fabrication process (there are engineering and cost considerations). Thus, there are three major classes of limits: (1) those imposed by the government for reasons of safety and spectrum allocation to reduce interference, (2) cost and size considerations based on the uses of the RFID system, and (3) engineering of the silicon and RF designs to make them more efficient in using power and lower in cost.

Each of these classes of limits can have a substantial impact on the design of RFID systems. For example, the operating frequency and bandwidth restrictions in different parts of the world mean that the same type of tag can be read at a rate of 500 tags per second in the United States but at a rate of only 200 tags per second in Europe. Thus, readers are limited by local standards because they generate RF emissions, whereas passive tags typically can work anywhere because the interrogating reader governs their backscatter transmissions.

Reader antennas can be quite sophisticated. Although some antennas try to capture all tags within a regular hemisphere, most are designed to have gain in a particular direction. Directionality enables the reader to focus its energy in a region of interest. The narrowness of the beam determines the angular accuracy. The speed of the identification protocol limits the number of tags that can be read per second and how quickly they can be moving.

Readers employ an arbitration protocol to identify each of a group of multiple tags sequentially. Various arbitration protocols are used in practice, including probabilistic slotted-Aloha, slotted-Aloha with random temporal backoff, deterministic binary-tree traversal, multi-bit deterministic tree traversal, and combinations of these protocols. Reader interference remains a challenge. Because, as described previously, reader signals decay as a square of the distance while passive tag returns decay as a quadratic, there is a fundamental problem in having high densities of readers. Thus, readers may ultimately require operating models that very carefully mitigate interference, just as must be done in multiple-access radio systems.

Readers for Tags with Memory and/or Sensing Capability

Besides reading more data from the tags, readers designed for tags with additional memory and/or sensing capability need to have additional capabilities themselves. One consideration is that of writing tag memory. Tags will most certainly not allow any reader to write into their memories (possibly destroying the data that were there). Thus, some security measures for tags, such as memory locking and password protection, must be in place. Some tags already have their memory arranged in sections with independent access control, so that one organization's readers can access one area while another organization uses a different area. How these access rights and passwords are managed is an important issue for the information systems that manage the readers.[6]

Another consideration in this area is that readers of sensor-enabled tags may be thought of as nodes in a sensor network. They can communicate with multiple tags within their operating range. The tags can provide many sensor readings that can be aggregated and forwarded through the sensor network. Thus, readers may take on the properties and requirements of sensor network nodes.

Readers for Active Tags

Readers for active tags have a range proportional to $1/d^2$ rather than $1/d^4$, because these tags have power of their own to use in transmission. To get twice the range between readers and active tags, power output must typically increase only by a factor of 4, rather than by a factor of 16 as for passive tags. Readers and active tags are thus much more comparable in their communication systems, as the communication needs are equivalent. Tags may use a lower transmitting power to conserve energy when communicating with other tags if their usage allows a shorter peer-to-peer range. Readers of active tags can also operate as nodes in a sensor network or simply provide one more communication hop as a gateway between a tag-based sensor network and the information infrastructure.

Today's reader costs are typically on the order of $100 to $1,000 for near-field readers reading passive tags and $1,000 to $2,000 for far-field readers reading active tags. Costs are likely to decrease rapidly, as the technology is likely to experience large economies of scale. (However, considerable uncertainty surrounds these future economies.) Antenna sizes can be quite small (1 centimeter) for near-field tags, but as large as 0.3 meter to 1.0 meter for some far-field technologies. Active tag readers have antennas comparable in size to those of tags—on the order of a few centimeters.

In summary, readers are the highly regulated elements of RFID systems. They must meet constraints imposed on their frequency of operation and their power output. Thus, they can be limited in the number of tags that they can read per unit of time and in the range at which they can communicate with tags. Active tags provide a way of boosting communication range, but at the same time they make tags more expensive (by requiring a battery) and possibly create a maintenance issue (battery replacement) if the lifetime of the tags needs to be longer than their period of use in a particular scenario. Readers carry out the crucially important function of connecting tags to the information infrastructure and thus to end-user applications.[7] How these data are communicated between sources and destinations can make the system quite complex, as the paths may often require connecting across network administrative domains. (This is especially the case in supply-chain management, where tagged packages move from manufacturer to transportation system to distribution center to a possibly different transportation system and, ultimately, to a retail outlet). Finally, readers may need to coordinate among themselves when their ranges overlap, so that they do not interfere with one another's abilities to communicate with tags.

---

[6] For more on RFID and security, see S. Sarma, A. Weis, and D. Engels, 2003, "Radio Frequency Identification: Security Risks and Challenges," *RSA Laboratories Cryptobytes*, 6(1), pp. 2-8, available online at <http://www.rsasecurity.com/rsalabs/cryptobytes/CryptoBytes_March_2003_lowres.pdf>, accessed December 14, 2004.

[7] In some cases, it might make sense to instrument the environment with a multihop sensor network of tag readers and to have the tags themselves do less. This arrangement would help increase the longevity and robustness of the tags. It could be extended further by accommodating mobile elements of infrastructure (that is, mobile readers that join, and leave, and rejoin more connected pieces of the reader-tag-sensor-network infrastructure).

COORDINATING CAPABILITIES AND SYSTEM CONSIDERATIONS

The day is rapidly approaching when every manufactured object could contain an RFID tag of some type. But just because this could be done does not mean that it would make sense to do so. The type of tag, if any, that is most appropriate for a particular use depends on many factors. It is important to reiterate that not all of the properties of tags discussed so far can be found in combination. For example, tags that do not have their own power sources cannot collect sensor data and communicate with other tags.

Figure 1.1 shows, along the horizontal axis of the graph, the range of assets that may require identification. These assets vary from an individual item on a store shelf to a trainload or airplaneload of material. Clearly, a single, higher-cost tag is more appropriate for a plane or truck. An active tag with Global Positioning System (GPS) capability is quite appropriate for these uses: it can be reused and easily maintained; it tags thousands of items as a group. At the other end of the scale, individual consumer items are likely to be tagged with bar codes for the foreseeable future. Even if a tag costs only a few cents, it is difficult to justify except for more expensive items with sizable profit margins. Of course, the value proposition may change in the future as new usage models emerge and consumers possibly become willing to pay the additional costs of tagged items.

Choosing a tag type does not depend solely on the value of the tagged asset. Properties of the reader-tag combination also play an important role. The read range and orientation sensitivity (the latter is sensitive to both the cost and design of the receiver circuit) are probably the most important concerns. Reader density is another important consideration. If fine-grain tracking is required, more readers need to be deployed at smaller intervals. One common case is that assets are tracked only at choke points, such as at warehouse loading docks or on truck beds. Other possibilities include tracking items at the level of a store shelf to enable automatic inventory control.

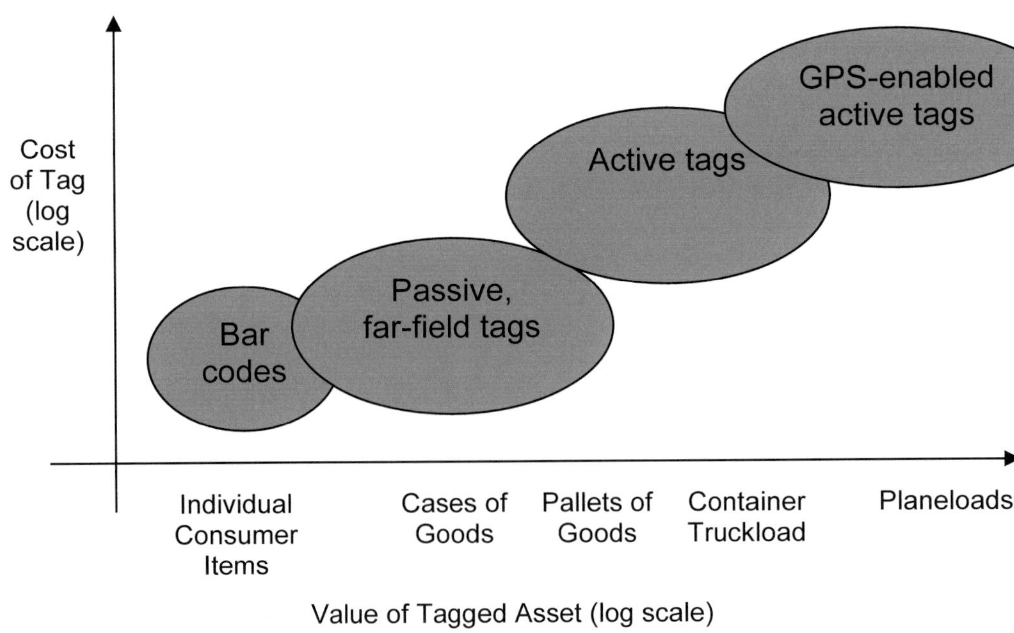

FIGURE 1.1 The types of tags appropriate for different types of assets. Both the vertical scale of tag cost and the horizontal scale of asset value are logarithmic.

What happens to tags as they move through the supply chain is also important. Some tags end up at home with consumers. Tag readers could continue to see such tags wherever customers take the items. It is possible that privacy considerations will require that tags be decommissioned at the retail counter so that they are no longer functional when customers leave the store. What happens to tags on cartons and cases? If not well managed, they could end up polluting the environment—not only physically, but also electromagnetically, by their signals.[8] Today, most tags used in the fast-moving consumer goods market have a password-protected kill feature that can permanently disable a tag.

Location is closely coupled with identification. Readers with GPS capabilities that can record their own locations and movements have already been mentioned. This type of reader is likely to be a very popular, although of course GPS capability increases cost.

An interesting adjacent technology for active tags is WiFi (a type of wireless network). The idea is to use WiFi (802.11x) access points as readers. These devices are rapidly gaining ubiquity as wireless networks spread around the planet. They can serve a dual purpose in gathering IDs from tags and reporting their approximate locations—within WiFi range. WiFi-based active tags are already beginning to be marketed commercially. However, as indicated by the focus of discussions at the workshop, and thus much of this report, far-field passive tags are garnering greater attention as they are nearing deployment in supply-chain management both for large retailers and governments.

STANDARDS BODIES AND STANDARDS

As described previously, RFID systems comprise readers and tags along with a back-end infrastructure for data management. Because RFID readers and tags communicate using electromagnetic waves, they are classified as radio systems. RFID systems do not generally have allocated frequency bands, but tend to use the unlicensed frequency ranges classified worldwide as ISM (Industrial-Scientific-Medical) or SRD (Short-Range Devices). As RFID systems become more prevalent, there may be a need to revisit the issue of dedicated spectrum for them, because large-scale RFID deployments may eventually monopolize the ISM or SRD bands.[9]

When contemplating RFID systems, it is important to realize that radio frequency regulatory requirements are not uniform worldwide. Consequently, RFID tags and readers that cross international boundaries must meet the "lowest common denominator" of competing national and international regulations.[10] RFID standards have traditionally been developed using

---

[8] One possible solution to tag pollution might be to design the reader signals to receive significant responses only from classes of ID sequences of interest (for example, clothing, not food), by having a taxonomy of product types. This could also partially mitigate some surveillance concerns: The reader would collect information only on classes of objects that it is authorized to scan. Other information would never enter the database.

[9] For more on the challenges of spectrum allocation and spectrum policy, see the Computer Science and Telecommunication Board's ongoing study on wireless technology prospects and policy options online at <http://cstb.org/project_wireless> and its associated workshop report: National Research Council, 2004, *Summary of a Forum on Spectrum Management Policy Reform*, The National Academies Press, Washington, D.C.

[10] The following documents are useful for understanding the range of worldwide regulatory requirements: ISO/IEC (International Organization for Standardization/International Electrotechnical Commission) 18000-1, "Air interface, Part 1—Generic parameters for air interface communications for globally accepted frequencies." ISO/IEC 18000-6, "Air interface, Part 6—Parameters for air interface communications at 860–930 MHz." ISO/IEC 3309, "Information technology—Telecommunications and information exchange between systems—High-level data link control (HDLC) procedures—Frame structure." ISO/IEC 19762-3, "Information technology AIDC techniques—Harmonized vocabulary—Part

the International Organization for Standardization process. For example, ISO developed the ISO 18000-6A and 18000-6B 900 megahertz (MHz) ultrahigh frequency (UHF) standards, which are used worldwide. In 1999, a joint industry–academic effort at MIT, the Auto-ID Center (for Automatic Identification) created two 900 MHz non-ISO protocols that have met with commercial success. EPCglobal, an outgrowth of EAN International and the Universal Code Council,[11] formally took over the commercialization of these protocols from the Auto-ID Center in October 2003. The Auto-ID Center was renamed the Auto-ID Labs and continues its academic effort in advancing RFID technology.

The two protocols are named EPCglobal Class-0 and Class-1 (see Box 1.1 for more information). Unfortunately, for technical and regulatory reasons these protocols are not suitable for adoption as international standards, so EPCglobal undertook to develop a single, worldwide UHF RFID standard, termed Class-1 Generation 2 (Gen2 for short) to replace them. The Gen2 standard was promoted to candidate specification on October 1, 2004, and is expected to be ratified by EPCglobal in December 2004. The SC31 subcommittee within the ISO has already announced that it will incorporate Gen2 into its existing 18000 structure as ISO/IEC 18000-6c as soon as Gen2 is formally ratified by EPCglobal.

The 900 MHz UHF band is rapidly emerging as the preferred RF band for supply-chain applications, primarily for reasons of read speed and range (see Box 1.2 for other RFID frequency bands and standards). Whereas passive 13.56 MHz tags can be read at rates from 10 to 100 tags per second and at a range measured in centimeters, passive 900 MHz UHF tags can be read at rates from 100 to 1,000 tags per second and at a range measured in meters. Specifications under development, such as EPCglobal's Gen2, will push the read rates above 1,000 tags per second and the read/write range beyond 10 meters. Further driving the adoption of 900 MHz RFID are the mandates from giant retailers Wal-Mart, and Target, from the Department of Defense, and from others that require suppliers to use RFID for tracking and inventory control starting in 2005.

The 900 MHz RFID is not, however, without limitation—indeed, 900 MHz RFID has far greater problems with signal fading due to multipath effects (interfering reflections of the signal that cause adjacent regions to vary dramatically in reception) and signal attenuation by liquids and metals than does 13.56 MHz RFID. But workshop participants observed that regardless of these issues, supply-chain applications are poised to accelerate the adoption of 900 MHz RFID, and with that acceleration, it is hoped, some solutions to these problems may emerge.[12]

3: Radio frequency identification (RFID)." U.S. Code of Federal Regulations (CFR), Title 47, Chapter I, Part 15, "Radio frequency devices, U.S. Federal Communications Commission." European Telecommunications Standards Institute (ETSI), EN 302 208, "Electromagnetic compatibility and radio spectrum matters (ERM)—Radio frequency identification equipment operating in the band 865 MHz to 868 MHz with power levels up to 2 W, Part 1—Technical characteristics and test methods." European Telecommunications Standards Institute (ETSI), EN 302 208, "Electromagnetic compatibility and radio spectrum matters (ERM)—Radio frequency identification equipment operating in the band 865 MHz to 868 MHz with power levels up to 2 W, Part 2—Harmonized EN under article 3.2 of the R&TTE directive." EPCglobal™ (2004) EPC™ Tag Data Standards Version 1.1, Rev. 1.24.

[11] See the Web site <http://www.ean-ucc.org/> for more information. Accessed December 17, 2004.

[12] It should be noted that most of the standards work under discussion addresses a range of technical specifications as described here. However, these standards do not necessarily address all possible usage models—such as data mining. Some kinds of uses involve more than just tags and readers—that is, they involve the entire system within which the RFID technology is embedded, resulting in vendor-specific implementations.

BOX 1.2
RFID Frequency Bands and Standards

The most common radio frequency identification (RFID) frequency bands and the standards associated with their usage are the following:

- **25 kHz HF—Near-field, all passive**
  —ISO 18000-2
- **13.56 MHz HF—Near-field, mostly passive**
  —ISO 18000-3 Mode 1, Mode 2
  —ISO 14443 Type A, Type B
  —ISO 15693
  —EPCglobal Class-1 HF
- **900 MHz UHF—Far-field, some active**
  —EPCglobal Gen2
  —EPCglobal Class-0, Class-1 (not standardized)
  —ISO 18000-6 Type A, Type B
- **2.45 GHz UHF—Far-field, some active**
  —ISO 18000-4 Mode 1, Mode 2

NOTE: HF denotes high frequency; UHF, ultra-high frequency; KHz, kilohertz; MHz, megahertz; GHz, gighertz. RFID systems also operate at other frequencies, including 433 MHz, 5.75 GHz, and others, but those that use the frequency ranges listed above are the most common.

In North America at least, RFID systems and their components have some unique features that are not shared by other users of the ISM band, including dense deployment (eventually hundreds of readers in a retail store), passive backscatter with spectral content usage across international boundaries (that is, tags must work in multiple international environments), and a need for the reader to transmit maximum power so that tags receive sufficient operating energy. Given these issues, sorting out how best to manage interference remains a challenge both for regulatory agencies such as the Federal Communications Commission (FCC) and for industry and standards groups. (See Box 1.3 for more on the FCC constraints for RFID operation in North America.)

APPLICATIONS AND BUSINESS

RFID technologies are being used in a wide variety of applications (see Box 1.4 for a sampling). Besides collecting tolls, probably the most common applications at present are in supply-chain management—monitoring goods as they move around a warehouse, through a factory, between distribution centers, and around the globe. It is for such purposes that Wal-Mart, Target, and the Department of Defense have mandated the use of RFID by their suppliers.

But participants in the workshop pointed out that the potential of this technology goes well beyond supply-chain uses. At a high level, RFID technologies can be viewed as a way to bring together the physical environment and the informational environment in many different contexts. Whereas in the past instrumentation has been possible, RFID allows the easy integration of instrumentation with communication. This combination is what presents such interesting possibilities to those experimenting with the technology in the business world. At the same time, understanding the capacities and limitations of the technology is critical to effective deployment. It is important to recognize the broad range of RFID technologies and to understand the application needs in order to determine what kind of RFID (or other) technology will work best in a given situation.

BOX 1.3

RFID in North America

The Federal Communications Commission (FCC) authorizes the 902–928 MHz frequency band for unlicensed Industrial-Scientific-Medical (ISM) devices. The following constraints must hold for all users of this band, including radio frequency identification (RFID) readers:

- The 902-928 MHz frequency band is unlicensed: Readers are authorized, rather than licensed, by the FCC for use of this band.
- Interference constraints are as follows: Readers may not cause harmful interference and must accept any interference received.
- Requirements specify the frequency band, power output, out-of-band and spurious emissions, and frequency stability. Examples include a specified maximum power output, channelization requirements, a maximum antenna gain, and so on.

Standardization efforts in RFID and adjacent technologies are needed to guarantee the private sector predictable deployments. Examples include cordless phones, baby monitors, garage door openers, personal digital assistants (PDAs), wireless local area networks (LANs), ultra wideband (UWB), contactless smart cards, and RFID.

The FCC periodically reviews its regulatory constraints on the basis of the following considerations:

- Public need and benefits for the service;
- Amount of spectrum required, considering technical limitations on spectrum efficiency and impact on economic viability of service;
- Controlling interference with other services;
- Other technical considerations, including the ability to control interference, radio frequency propagation, apparatus limitations; and
- International allocation considerations, including use in neighboring countries (primarily Canada and Mexico) and the need for international harmonization of service.

This is a dynamic moment in the life cycle of this technology. Anything and everything seems possible. The cost and capabilities of the various ID technologies—bar codes as well as RFID—continue to improve. The standards process, despite rapid progress, is as yet incomplete. Thus, it is unclear which types of RFID technology will prevail. Many companies are experimenting, but the results are highly proprietary. In place of well-accepted data, anecdotes abound. A current challenge is that much of what is being done with RFID technologies is in a closed-system context—that is, within a company, organization, or small group of organizations. This results in proprietary information that makes it difficult to provide general knowledge about the technology and systems challenges that could help advance understanding within the broader community. In sum, it is very difficult now for a single individual or firm to evaluate the costs and benefits of the technology, even for a well-defined application such as supply-chain management, and still more difficult to evaluate speculative future applications.[13]

[13] Larry Dignan and Kim S. Nash. 2004. "RFID: Hit or Myth?" *eWeek*. February 9. Available online at <http://www.eweek.com/article2/0,1759,1524634,00.asp>. Accessed December 14, 2004.

BOX 1.4
Sample Applications of RFID Technologies

Applications of radio frequency identification (RFID) technology are proliferating rapidly. The following list presents a sampling of RFID applications; such a list could not be exhaustive as new uses are reported almost daily. Each application presents technical and social challenges.

- **Supply-chain tracking** is the most discussed RFID technology application, given mandates by Wal-Mart, the Department of Defense, and many other retailers and manufacturers that suppliers must provide pallet-level RFID tags for tracking goods through distribution networks. The objective is to minimize losses due to logistical failures and/or theft. Tags need to be read in large quantities at speeds that do not interfere with work practices.
- **Homeland security** is a natural arena for the application of tracking technologies. Assuring the security of container shipping is of particular concern. Tamper-detecting active RFID tags on containers as well as passive tags on items within containers are being proposed as a way to streamline port operations while increasing security.
- **Livestock tagging** is likely to become more important as outbreaks such as of mad cow disease continue to spread globally. Recent incidents of infected cows have demonstrated the importance of being able to track the movement of animal products through a series of facilities on their way to market. Current record-keeping techniques have not proved adequate, thereby triggering the search for new solutions.
- Many **pets** in the United States and elsewhere are implanted with subcutaneous IDs so that if they are ever lost, they can be traced back to their owners.
- The State Department is considering the possibility of issuing **passports** that include biometric information encoded on RFID tags. Trials using passive RFID and contactless smartcards are in progress. Concerns about privacy are very much at the forefront. Citizens carrying passports with encoded information could be tracked or identified by tag readers belonging to entities other than the federal government. Encryption of ID data will be a key issue as well as what information is actually stored on the tag versus what is in an associated database (for example, border entry/exit points crossed by a passport).
- The Justice Department of Mexico is implanting RFID tags in some of its **employees** with access to sensitive information. Implantees and the documents they access are logged when these employees enter data vaults.
- A **nightclub** in Barcelona seeking to provide patrons with extra convenience is implanting RFIDs so that its customers do not have to carry wallets or purses. Proper feedback and control mechanisms are needed to ensure that charges are properly assessed.
- **Tracking pharmaceuticals** from large containers in the supply chain down to individual doses may be effective in preventing counterfeiting, dilution, and theft. By observing the entire supply chain, losses can be identified where and when they occur. At the consumer level, tags on individual packages can help identify potential drug interactions in the home as well as in hospitals.
- E-ZPass and FasTrak systems are used in some regions of the United States to collect **highway and bridge tolls**. These applications primarily use semiactive tags that can boost the signal back to the reader, thereby increasing the range and speed of reading so that drivers do not have to slow down as much as they would otherwise. There are many other suggestions for **electronic payment systems**, ranging from toll collection to electronic wallets implemented as contactless smart cards. Nokia has introduced an RFID tag reader in a phone that can be used to read the tags on a vending machine for automating the purchase after pointing the short-range reader at the product's tag.

- A few Japanese primary schools are **tagging students' belongings** (for example, clothing, bags, and name-tags), so that teachers and parents can track the children's whereabouts. Readers at key positions around the school can provide information about when and where students were last seen or whether they are within the bounds of the location of their expected activities.
- **Theme parks** are beginning to use RFID technology to provide reservation services for rides and the tracking of family members to enable them to find one another more easily via kiosks around the parks.
- **Marathon** organizers have used RFID tags for several years to track the positions of runners throughout the race course. This information can be shared with fans so that they can follow the progress of the athletes individually or as a group.
- **Museums** are using RFID tags in several ways—most obviously to track and identify artifacts and as a security measure. More interestingly, the San Francisco Exploratorium, a hands-on science museum, is using RFIDs given to visitors to aid them in collecting information of interest onto a personalized Web page that can then be enhanced with other information and activities that relate to those topics in an attempt to lengthen engagement with the exploration begun at the museum.
- Researchers in **ubiquitous computing** are using RFID tags as a way of tracking people and objects. Applications range from a reminder system that generates alerts when people leave objects behind to data gathering about elderly people's activities in the home so that caregivers are better informed and can analyze trends that may indicate cognitive or psychological decline.
- **Luggage tracking** is another area in which several experimental studies are ongoing. Bar code systems typically have too high a miss rate for this use owing to line-of-sight requirements and orientation requiring frequent human intervention. RFIDs in luggage tags would provide more accurate and more automated routing. Results so far have been mixed, as tag orientation is still affecting read reliability at too high a rate.
- Using **RFID in clothing** was seriously set back when plans of clothing retailer Benetton raised serious privacy concerns over how the consumer would control the tags. Applications in this area range from receiptless transactions (the receipt is in the tag) to fashion advice given a set of clothing items (what matches and doesn't, suggestions for purchases), to automatically determining washing machine and clothes dryer settings.

The Supply Chain and Beyond

Supply-chain management has been one of the driving business applications behind RFID. Companies such as Wal-Mart are mandating the use of RFID,[14] thus providing incentives for smaller companies to begin incorporating the technology into their inventory management and supply-chain systems. The sheer volume of materials and products moving around the country and the globe on any given day is a powerful motivator for finding increased efficiencies. Much of the focus today is on decentralized supply chains, in which many different organizations are involved in a geographically distributed effort. Workshop participants noted that RFID technologies have the potential to improve the capacity of organizations to "trust but verify" when dealing with partners.

RFID is suggested to add value to several components of the supply chain. At a high level, it can assist in tracking product flows and transmitting demand signals back up the supply

[14] Workshop participants reported that Wal-Mart plans to require EPC tags at the case and pallet level from dozens of its top suppliers by early 2005 and from all of its suppliers in 2007. See also Mark Roberti 2003, "Analysis: RFID—Wal-Mart's Network Effect," *CIO Insight*, September 15, available online at <http://www.cioinsight.com/article2/0,1397,1455103,00.asp>, accessed December 14, 2004.

chain faster than other methods can, resulting in shorter replenishment cycles. More specifically, RFID systems and the information that they generate can assist in the improved structuring of warehouses, location of goods, optimal production and distribution batches, and so forth. They can also be helpful in ensuring a "first-in, first-out" process when the product being tracked is perishable. They could allow improved monitoring for spoilage, coordination with thermal tags and expiration data, proof of delivery, and various other kinds of information associated with products moving within a supply chain. (Such possibilities, of course, require much more capability than that of simple passive tags.) In general, RFID systems have the potential to provide a vast amount of information on business processes, but it is up to the companies to sort out how to make that information useful.

It was reported at the workshop that at this time several pilot studies and experiments are being run to determine how best to incorporate RFID technologies into supply-chain management strategies. In terms of consumer products, one question that arises is whether it makes sense to place tags on single items, or whether the tags (and associated tracking) should be kept at the pallet or case level. It may turn out to be most cost-effective if only high-value items are tagged (for example, cases full of many individual objects, or large single high-value items such as DVD players). Workshop participants suggested that in many cases pallet scanning rather than item scanning was most effective.

Another issue that arises with respect to consumer products is whether and how to move to a fully RFID-enabled supply chain. One participant observed that Wal-Mart, for example, will most likely have a hybrid system (employing bar codes along with RFID) for years. Understanding the implications of that type of system will be important to moving forward.

For a supplier to a huge customer that requires RFID, there is little difficulty in deciding whether and what kind of RFID to use. For others, however, the problems can be vexing. For example, the idea of continuous, real-time inventory monitoring is an exciting prospect. That sort of application, however, goes beyond the capabilities of the simplest and cheapest RFID tags and readers. One workshop participant described such an application, but the objects tagged were large shipping containers. At that level, more expensive and thus more capable tags and readers may make sense. The dream of continuous inventory monitoring at the item level may for now be just that, a dream.

While supply-chain management is an obvious and noteworthy application of RFID technologies, workshop participants described many other potential applications that businesses might consider. In addition to object tracking within a supply chain, other, related applications include reverse logistics (e.g., tracking returns), quality management, marketing, inventory, accounting, assistance in allocating overhead costs, warranty tracking, and recycling. For example, it was suggested that RFID technologies could assist with quality management by providing information about which items have gone through a problematic section of a production process. Such technologies might also aid in delaying product differentiation—that is, in enabling a large variety of products to be made from common materials and components until feedback via RFID helped determine specifics of needed products. This kind of production process requires extensive communication systems in order to decide exactly which mix of products to make, and RFID could well be a valuable part of such systems. Similarly, by tracking parts and components more precisely, RFID might enable more accurate cost accounting. These and other sorts of applications will all have strategic implications for organizations.

Workshop participants also suggested that RFID technologies could assist in schedule optimization, not just in the supply chain, but in the dispatch of service and delivery vehicles as well as in more efficient deployments (and redeployments) of emergency assets. Again, this type of application relies on more precise information about the location and status of these assets.

Health care and security were also discussed as potential application domains. In health care, virtually every component—including patients, doctors, equipment, and drugs—could be tagged, tracked, and monitored.[15] Closer and more sophisticated supply-chain monitoring enabled by RFID could help prevent drug counterfeiting or tampering and reduce spoilage. Tags on ID cards could help prevent unauthorized entry into restricted areas, and tags on objects could help prevent theft. (It was also pointed out, however, that ill-intentioned people could likely devise ways around such straightforward security measures, for example by spoofing tags. Therefore, more sophisticated measures that change the behavior of the system over time might be necessary.[16])

In summary, as seen in the discussion of tags, readers, and architectures, the term *RFID technology* refers to a broad spectrum of functionalities. RFID technologies can do many things. In a business context, it will be important to identify specific objectives and to understand and select the subset of RFID technologies that offer the needed capabilities.

Consumer-Centered Applications

Discussions at the workshop pointed to a major focus by business on the supply-chain and related RFID issues, while much attention from consumers has been focused on the retail experience. Participants reported that there has been some activity in retail, although not as much as in the supply chain, but that it is harder to build a compelling business case for widespread RFID deployment in a retail environment at this time. The idea of eliminating checkout lines may be appealing, but for now the cost seems prohibitive, and various privacy and security issues remain to be resolved. Nor was it clear whether significant advantages would accrue to consumers or businesses over what current bar codes and handheld scanners offer.

An experiment conducted at an upscale clothing store was reported on. For this experiment all items of clothing were tagged, and kiosks presenting various kinds of information about the products were available for sales staff to consult when working with customers. The results were mixed. Although the sales staff appreciated the ready access to information, they disliked the distractions of operating the system, preferring to focus all their attention on the customers. Also, the company was unable to take advantage of other possible uses of the technology. For example, it did not have the information infrastructure to interpret the frequency with which items were scanned and turn those data into useful information for making inventory and stocking decisions.

This experiment and other reports from workshop participants raised several points to be borne in mind when the deployment of RFID technologies in a retail environment is being considered, if the goal is to do more than provide a faster checkout at the point of sale. It is important to focus carefully on what the retailer and its processes require. Depending on the details of the deployment, RFID technologies and their associated systems could in effect provide too much (or even inappropriate or useless) information, to both sales staff and customers. (See Box 1.5 for more on information overload and RFIDs.) This possibility suggests that careful design, along with good interfaces and effective information management, will be important.[17]

---

[15] Jonathan Collins. 2004. "Hospitals Get Healthy Dose of RFID." *RFID Journal*. April 27. Available online at <http://www.rfidjournal.com/article/articleview/920/1/1/>, accessed December 14, 2004.

[16] As with almost all information technology systems, RFIDs will pose security challenges. For more on computer and system security, see the following publications from the National Research Council: *Trust in Cyberspace* (1999), *Cybersecurity Today and Tomorrow: Pay Now or Pay Later* (2002), *Cryptography's Role in Securing the Information Society* (1996), and *Computers at Risk* (1991).

[17] For more on the challenges of managing complexity in technology, see the recent survey of information technology from *The Economist*: "Keep it Simple," October 28, 2004, available online at <http://economist.com/surveys/showsurvey.cfm?issue=20041030>, accessed December 14, 2004.

BOX 1.5
Information Overload and Data Mining

One of the issues that came up in many contexts at the Radio Frequencies Identification (RFID) Workshop was information overload associated with RFID system deployments. A purported advantage of RFID technology is that it provides increased information visibility along with improved information flow. However, as the retail experiment that was described at the workshop demonstrated (see the subsection "Consumer-Centered Applications"), this can mean that too much (or unhelpful) information is presented to the end user without an effective way to process, manage, or use it. An increased amount of data, by itself, will not necessarily improve business, so carefully constructing a business case is important. Organizations need to know what they are measuring and why. One of the biggest challenges is determining how to collect the right information and how to provide it at the right time in order to support good business decisions. To do this, the information must be collected well, stored well, and then presented well to the user (be that user a system manager or a customer). Collecting data for its own sake is unlikely to be useful and may often be distracting.

The issue of data mining with respect to RFID-generated data was also raised at the workshop. Participants noted that while this topic gets a lot of attention when people talk about RFIDs (and related information technologies), there was not, in their experience, a compelling business case for extensive data mining yet. Demand for data mining would also imply significant changes in the supply chain beyond what RFID for item tracking and some of the other applications require. In a retail context, for example, it was noted that it is not at all clear what it would mean for a business to know that a particular item was picked up and then placed back on the shelf. It is also not clear how much benefit the average retail operation could derive from RFIDs for item tracking beyond what bar codes currently provide.

Sophisticated data mining and data organization would seem to be a solution to the challenge of information overload. But both are hard technical problems beyond the scope of the workshop. Similarly, collection of data, especially data identifying individuals, raises obvious privacy concerns (discussed in Chapter 2). RFID technologies pose new challenges in several areas—appropriately and effectively managing the data produced, along with choosing what data to collect, are yet two more.[1]

[1] For more on managing RFID-related data, see Robert Whiting 2004, "Data Avalanche," InformationWeek, February 16. Available online at <http://www.informationweek.com/showArticle.jhtml?articleID=17700027>. Accessed December 14, 2004.

Also mentioned at the workshop was another application area besides traditional retail sales involving direct consumer interaction with tagged merchandise—that of contactless smart cards for personal finance. It was observed that it is still too early to tell what all of the possible applications might be with respect to financial transactions or in retail environments beyond the supply chain and beyond tracking merchandise (and information about merchandise). There are RFID pilot deployments underway in many different organizations, even museums. It will take time, unfortunately, for the results of these pilot studies to be useful to the public and in various enterprises. There is a high cost in doing these studies, and their results could provide a competitive advantage that an organization may not want to share so readily.

As with many basic, multiuse technologies, it is very likely that innovation in RFID technology will continue and new applications will emerge over time. All such applications, however, will likely raise the social and cultural issues discussed in Chapter 2.

Ongoing Challenges

Throughout the discussion of business needs and applications for RFID technologies, workshop participants noted several persistent technical and policy challenges. At a general level, the stability of the technology and the associated standards will be significant factors influencing the business case regarding RFID. In addition, the Wal-Mart, Department of Defense, and other mandates will have an impact on who uses this technology as well as on how it is used. Participants suggested that making the business case for companies not heavily influenced by those huge organizations will be critical if the technology is to see broader deployment. Some smaller companies may look to RFID as a way to move past (or skip) the use of bar codes, but much affecting whether this might be possible is still at a very experimental stage. Other drivers of regulatory change aside from Department of Defense requirements include those of various governments to provide country-of-origin food labeling, pharmaceutical tracking, other asset tracking, and techniques to prevent counterfeiting. The Department of Homeland Security is also taking an interest in using RFID technologies to secure shipping containers against tampering. Interestingly, depending on the requirements from any of these organizations, active tags may be more likely to be subject to regulation than passive tags.

Although some press accounts would seem to suggest that RFID technology is very simple, in fact there are complications that make large-scale deployments a challenge. For example, choosing where to place the tags on an item is a serious issue for some applications. Most tags cannot be read through liquid or on cans. In addition, while the tags are relatively inexpensive, readers are not. Database and infrastructure requirements also add to the cost of implementation. Typically, it was noted, RFID technology costs can be thought of as roughly evenly divisible between software, hardware, and systems integration. Some workshop participants suggested that the business case has not justified the cost of the technology in many arenas and that experiments with RFID systems are still in their infancy.

2

Society and Culture

In an experiment at the San Francisco Exploratorium, a hands-on science museum, families are given a handheld radio frequency identification (RFID) reader and assigned a unique Web address that will document their visit to the Exploratorium. As they move through the various exhibits looking at different parts of the displays, they can use the reader to query tags posted throughout the museum, and the system logs their interests. When they get home, the particular exhibits that they visited come up on the Web site created for them, linking to additional educational material to spark further interest. The same handheld reader is being developed for senior citizens to help with elder care.[1] In manufacturing contexts, RFID systems in warehouses track inventories of goods with the aim of substantially reducing theft and loss while increasing efficiency, and thus potentially reducing the cost of goods to consumers.

Given these sorts of informative, helpful, and cost-reducing applications, it is interesting to note that more than 40 of the best-known European and U.S. consumer, privacy, and civil liberties organizations endorsed a moratorium on RFID tags applied to consumer products.[2] In addition, retailers Benetton and Wal-Mart both halted their early in-store tests of RFID inventory control systems—Benetton was even threatened with a boycott, and for both companies tangible economic benefits were not immediately obvious. While the potential benefits of the technology are vast, there are risks inherent in large-scale deployment of RFID. Until stakeholders (including industry leaders, policy makers, and advocates) grapple effectively with those concerns, it seems likely that interest groups will seek alternative means to make their voices heard. Strikes, boycotts, and protests have already been organized to effectively block RFID implementations.[3] This chapter provides a brief overview of many of the ethical, legal, cultural, and social issues related to RFID technology, drawing on discussions and presentations at the workshop.

---

[1] This is the iReader, developed by Intel. For a brief description, see Adam Rea, Waylon Brunette, and Gaetano Borriello, 2004, "Designing for Flexibility: A Look at the iReader," presented at the Second International Conference on Pervasive Computing. On the topic of elder care, see Celeste Biever, 2004, "RFID Chips Watch Grandma Brush Teeth," March 17, NewScientist.com News Service, March 17, 2004, available online at < http://www.newscientist.com/article.ns?id=dn4788 >, accessed December 14, 2004.

[2] "RFID Position Statement of Consumer Privacy and Civil Liberties Organizations," November 20, 2003, available online at <http://www.privacyrights.org/ar/RFIDposition.htm>, accessed December 14, 2004. A moratorium is being called for until a formal technology assessment, with substantial public participation, takes place.

[3] The 2002 longshoremen's strike, a boycott threatened against Benetton, and other protests against RFID technologies are mentioned in the sidebar entitled "Dealing with the Darker Side" in Roy Want, 2004, "RFID: A Key to Automating Everything," *Scientific American*, January, pp. 56-65.

The so-called "internet of things" enabled by RFID systems conceptually "make[s] it possible for computers to identify any object anywhere in the world instantly."[4] Such a vision holds tremendous promise in contexts such as inventory management and shipping and handling, as well as in hospital care, education, and safety monitoring. But, clearly, the promise is burdened by equally tremendous possibilities for misuse. The RFID Position Statement mentioned above (see footnote 2) lists five potential threats to privacy and civil liberties from the large-scale deployment of RFID technologies: hidden placement of tags, unique identifiers for all objects, massive data aggregation, hidden readers, and individual tracking and profiling.

As described previously, "Big Brother" scenarios in which commercial interests or government can track an individual's every purchase and move by compiling vast quantities of minute data from electronic product codes within RFID tags are some time away from being realized. But possibilities for immediate misuse remain. Workshop participants argued that addressing social, ethical, legal, and cultural concerns is crucial for RFID technologies at every stage, including technological design and the development of industry standards, policy and regulation, and specific applications. Developing policy to incorporate social norms in emerging technologies is often a discouragingly long process in an industry that seems to move at lightning speed, almost always ahead of the policy discussions as well as ahead of the purview of government regulators. For that reason, this chapter on the societal and privacy concerns associated with RFID technologies necessarily takes a long view of the technology and its potential implications.

Particularly in the realms of social norms, ethics, and policy, RFID technologies confound simple discussion in a number of ways. First, the differences in the speed at which policy and technology develop forces policy into what would seem to be the realm of science fiction. If policy discussions are forced to make assumptions about future technological developments, policy may fail to fit appropriately with societal interests as they evolve along with the technology. Second, because of what is often referred to as "function creep," technologies designed for one task are often adapted to accomplish another. Thus the stated purpose of any new system will be an incomplete description of that system's eventual use. Third, two primary means exist for incorporating social goals (be they privacy, security, manageability, reliability, or usability) into a system. The two means are regulation and design. Each is very differently motivated into action. Fourth and finally, thus far the most articulate discussion of social goals related to RFID technologies centers on notions of privacy. Privacy, however, is not universally defined, nor is it a flexible enough concept to encompass all of the issues that must be taken into consideration as RFID-incorporating systems become prevalent.[5] And, of course, privacy is far from a homogeneous or single-dimension concept.[6] Not only must the complexity of privacy as a concept be recognized, but other (sometimes related) concerns should be explicitly articulated as well.

Now is a good time for a thoughtful consideration of societal, cultural, and ethical issues related to RFID systems. A brief workshop cannot do justice to the complexity of all these

---

[4] "The Internet of Things" was the title of a Forbes article in March of 2002 by Chana R. Schoenberger <http://www.forbes.com/global/2002/0318/092.html>. Accessed December 14, 2004. (The quote is taken from <http://archive.epcglobalinc.org/aboutthecenter.asp>, accessed December 14, 2004.) The AutoID center has been incorporated into EPCglobal, and archives of the former organization should be available at <http://www.epcglobalinc.org/>, accessed December 14, 2004.

[5] One thing that makes "privacy" particularly challenging is that it has a weak feedback loop—it is not always immediately obvious when privacy has been affected or violated.

[6] Individual thresholds vary with respect to privacy—what one person might consider deeply private, another might casually disclose. Moreover, some argue that privacy encompasses more than merely the concealment or revelation of information, also being connected to autonomy and trust. Privacy is also deeply tied to context—behavior in one circumstance may be considered much more acceptable from a privacy standpoint than that same behavior in another situation.

aspects of RFID technologies, but discussions did present some of the basic issues and challenges. An early step is to identify stakeholders—both those who interact directly with the technology and those whose lives may be affected by it. The technology is currently rolling out, and will continue to evolve in the coming years, becoming much less expensive and presumably gaining wider application and currency. At present, development is driven, as noted in Chapter 1, by supply-chain market forces. While wholesale pallets of goods and designer shoes may be tagged for such purposes as managing inventory and controlling forgery, few consumers are piqued by these applications. But both function creep and speed of development will push seemingly neutral applications into the public sphere of debate. Some workshop participants argued that, to address public concerns most effectively, it makes sensc to develop and deploy RFID technology—even in these first, seemingly relatively neutral domains—with social norms in mind.

CHARACTERISTICS OF RADIO FREQUENCY IDENTIFICATION TECHNOLOGY IN SOCIAL TERMS

RFID technology poses interesting challenges in terms of its cultural significance because of the following interrelated features: the tags are minuscule, and some do not require their own energy sources; the systems are mobile and potentially invisible; the readers operate wirelessly at relatively close range; and components (especially tags) are cheap.[7] Altogether, these features characterize a technology that is likely to have broad applications and distinctive conceptual and cultural implications. For example, tags on or near individuals can be read without their knowledge, the tags can potentially be on everything in an individual's possession, and individuals could carry a complex, unique constellation of data (that is, of the full set of IDs on an individual's person or in his or her possession). RFID systems enable at least a scaling up, if not a change in the nature of surveillance and in the character of information collection that is possible.

Often discussion focuses on the tags, but the reader is a significant component of the system both technologically and with respect to social implications. In their most rudimentary implementation, the tags are passive whereas the readers are active agents. Thus, the tag and tag-bearer (be that a can of beans or a person) are in an asymmetric relationship with the reader or power source. The latter's agent has the capacity to interrogate the pervasive tags, and this imbalance means that readers, which can themselves remain hidden, will be able to gather information with relative impunity. The relatively short range at which tags currently must be interrogated will be extended to some degree as the technology improves. The fact that tags currently can identify themselves to any reader that supplies them with power means that data compiling can occur in unintended or unanticipated ways. It is important not to lose track of the reader in discussions about RFID technologies.

The specific characteristics of RFID technology have social implications. The most basic form of RFID tag is the passive tag that, in EPCglobal's taxonomy, for example, carries only its unique Universal Product Code (UPC)-like identification code: its electronic product code (EPC). The technology is poised at the head of several likely paths. Each makes clear that there are even

---

[7] Sensor networks (of which RFIDs could be considered a rudimentary version) also exhibit a distinctive collection of characteristics and operate under unique constraints. The challenge is to build large systems that are tightly coupled to the physical world and to one another in a resource-constrained environment that will persist for long periods of time while consisting of many interacting components and being used and interacted with by nonexpert users. The Computer Science and Telecommunications Board's 2001 report, *Embedded, Everywhere: A Research Agenda for Networked Systems of Embedded Computers* (Washington, D.C.: National Academy Press), articulates some of the technical, research, and social challenges presented by such systems.

greater cultural implications as technological capacity is increased. The implications include the following: (1) Passive tags are already being designed to contain more than an ID code. There will be increasing storage capacity for other types of data. (2) The tags will shift from read-only to read-write. (3) They will move from being passive tags to behaving as active nodes in a sensor net.

With each of these shifts, the privacy-related questions are compounded, and ethical concerns grow. The basic questions move from "Will I tell you my ID?" to "Will I let you write my data?" to "Will I let you tell me what to do?" In this chain of events, the reader's functional advantage over the tag becomes more serious. What ethicists and privacy advocates worry about is that each step, path determinism will occur—that is, current developments will determine certain subsequent outcomes and preclude other evolutionary alternatives. For example, when tags store more data in their next iteration, will they inevitably forgo privacy protections that will be needed, say, 3 years from now?

At present, the incipient application driving RFID evolution is the supply chain and to a lesser extent, military applications. Producers and vendors will tag products so that they can be read throughout the supply-chain process, from the production source possibly all the way to the retail outlet. RFID-enabled systems have the potential to reduce error, loss, waste, and theft in the supply chain, thereby increasing efficiency and profits. In this early stage of the evolution of the technology, tags report their IDs from pallets or cartons, and in-shelf readers can monitor stock inventory. At the same time, relatively expensive individual products are tagged at designer clothing stores or, for example, on the cartons of home entertainment systems. Later, when the cost becomes low enough, all products could be tagged. Analogous to UPCs, the RFID tags on pallets and shelves have limited implications with respect to cultural and social values.

Once tags are embedded in individual products, privacy and trust concerns become much more salient. Similarly, the location where tags are queried makes a difference. Tagged consumer goods in warehouses raise few social issues (although the dockworkers' strike on the West Coast in 2002 against tagged pallets demonstrated that the fear of lost jobs and on-site worker privacy could be significant concerns). In-store tracking of tags holds more social implications than the tagging of goods in warehouses, although in-store tracking, too, might be manageable through thoughtful policy development. When tags remain active after leaving the store, however, the social implications are potentially vast.

When thinking through these issues, it is crucial to understand when individual identity becomes associated with a tag (or data within the system), whether that needs to happen, and what the implications are. The capacity for data cross-referencing and linkage, though, means that even without explicit links to personal identity, connections to individuals will be possible. A current example, the prospect of RFID tags in car tires, serves to elucidate some of these issues. The scenarios begin to unfold regarding information that can be collected and tied to a unique individual, even if only passive tags are embedded in the tires and those tags are not associated with any particular individual. Commercial interests could canvass sports arena parking lots and market targeted products when those same tires park at regional shopping malls. Law enforcement could document cars parked at a political rally or at a rave where drug use is anticipated. Data logged at gas stations could be queried to track suspected criminals through the use of their cars. With read-write tags, data logs containing information about location, time/date, and service could be kept in the tire's tag memory. Particular products could be linked to unique personal identification—say, linkage of the tire's EPC with credit card data at the point of sale—enabling clear linkage of data and potential invasions of privacy. But even without the link to a specific individual, RFID technology poses potential problems due to the possibility of unintended, unauthorized, and undesired uses. Absent some type of notice, those negative uses can transpire without a consumer's knowledge.

TYPOLOGY OF RISK—PRIMARY SOCIAL CONCERNS

Beyond privacy, there are many contexts in which RFID and related technologies will have social and cultural impacts. At the extremes, commentators' narratives of those contexts describe either a utopian view of the future, in which emerging technology plays an enabling and empowering role, or a dystopian one, in which technologies fundamentally bring about civilization's demise. Mark Weiser's oft-cited prediction that ubiquitous computing "will make using a computer as refreshing as taking a walk in the woods" portrays an optimistic future.[8] By contrast, some privacy activists tend to believe that they must demonstrate what they believe to be the otherwise underestimated negative implications of technologies. Instead of either extreme, what is called for, as was discussed by participants at the workshop, is critical analysis and debate so that both risks and benefits can be explored.

Most importantly, those risks and benefits must be explored before the technology is fixed—that is, before the technology has been fully designed and developed for its various, specific applications. There is likely to be tremendous benefit in a design approach that precedes any sort of legislative or regulatory solution, not only because the resultant technology will almost certainly be more elegant, but, more importantly, because the public trust would not have been undermined. Should cultural concerns (privacy, security, legality, equity, and so on) be inadequately accommodated, a backlash of some sort is more likely to occur. Moreover, if lack of attention to privacy and social concerns means that advocates are forced to be more confrontational in order to have their concerns heard, it is possible that socially constructive uses of RFID technologies, from education and medicine to commercial applications, will be stymied.

Because social norms evolve, the technologies and the regulations for RFID technology must be agile. While some privacy advocates want government standards and public policy to regulate these technologies, the Federal Communications Commission (FCC) takes a relatively hands-off stance (see Box 1.3 in Chapter 1). The FCC sets minimal standards primarily to control interference, with the intention that broad, flexible rules will not only encourage innovation but also spur private industry to develop its own standards. This approach seems to make more sense for standards that promote technical and functional interoperability than for standards that seek to reflect social norms. Many argue that society's interests are best represented through public processes that include representation by various stakeholders.

In taking up the challenge of addressing privacy issues, EPCglobal has set forth in the Generation 2 (Gen2) standard, which comprises global standards for RFID, not only commercial guidelines to ensure uniformity, but also a basic response to privacy concerns.[9] The Gen2 standard establishes the basic tag as one with a 32-bit "kill" password, along with access control. The kill feature permanently disables the tag, acknowledges when the kill has been performed, and then goes silent.[10] According to the Gen2 standard, access-controlled tags cannot be written until the reader supplies the correct 32-bit password. Public concerns remain about whether the passwords could be intercepted or the tags broken into in some way. In addition, imposing authentication requirements on a technology raises a host of new technical and privacy

---

[8] Mark Weiser 1991. "The Computer for the 21st Century," *Scientific American*, September, pp. 94-104. For the Web version, see <http://www.ubiq.com/hypertext/weiser/SciAmDraft3.html>, accessed December 14, 2004.

[9] See the press release dated May 13, 2004, from Matrics, Inc., for example, in "Industry Leaders Propose EPC UHF Gen 2 RFID Standard," available online at <http://www.matrics.com/news/releases/20040513.shtml>, accessed December 14, 2004.

[10] Of course, for this feature to be effective, it will be important for users to be able to verify that a tag has been killed. Potential solutions for enabling this verification include home readers, public kiosks, or some sort of certification of trusted vendors, among other things—such solutions of course raise a host of additional infrastructural and social challenges.

challenges.[11] Other design options might be considered. For example, a system could be designed to limit the disclosure of data, depending on who (which reader) is requesting it, but again this solution requires a more sophisticated infrastructure than the most basic passive tags could accommodate.

It is impossible to lay out the exact components of a socially acceptable RFID technology, in part because the public has multiple interests and in part because RFID technologies have not undergone thorough assessment from a policy standpoint. Indeed, the primary recommendation of advocacy groups is to conduct a formal technology assessment of RFID. At present, the Fair Information Practice Principles should be noted, along with the privacy guidelines of the Organization for Economic Cooperation and Development.[12] A minimum set of guidelines culled from these sources yields five basic provisions for a possible RFID policy:

- *Notice/awareness*—transparency in the use and maintenance of RFID systems, with clear labeling, and without secret databases or tag reading; visibility;
- *Purpose specification*—notification of the purpose of any tag or reader;
- *Collection limitations*—collection of information limited to the purpose at hand;
- *Accountability*—RFID users responsible for complying with privacy provisions; formal entities to be established for monitoring and complaints; and
- *Security safeguards*—verifiable security and integrity in transmission, databases, and system access.[13]

In addition, a number of technical strategies are being explored to begin to address these principles. Kill switches in tags to be activated at the point of sale are the most obvious. Some advocate that RFID tags be permanently deactivated before being taken out of the store. Blocker tags provide another possibly strategy: Consumers would mask the transmission of any RFID tag in their possession through the interference generated by a blocker tag.[14] Features that enable anonymity are also useful in some contexts. Fundamental to the success of each of these strategies is the principle of notice/awareness: Users must be able to see that a tag is deactivated, has been interrogated, is concealed, or is active. While each strategy has particular utility, none is fail-safe and thus does not fully address the range of concerns expressed by consumer groups.

Another debate about ethical, legal, and social implications of RFID concerns consumer choice. Many civil libertarians would grant individual choice under all circumstances. Because these technologies can be complicated and their use coerced (through incentive programs, for example), however, it is hypothesized that two groups will form among consumers: those who want the ability to kill the RFID tags on their products and care enough to undertake the required actions, and those who do not. One of the principal concerns about RFID is that it will be difficult

---

[11] See the following Computer Science and Telecommunications Board report: National Research Council, 2003, *Who Goes There? Authentication Through The Lens of Privacy*, The National Academies Press, Washington, D.C.

[12] The Federal Trade Commission's articulation of Fair Information Practice Principles can be viewed online at <http://www3.ftc.gov/reports/privacy3/fairinfo.htm>, accessed December 14, 2004. The 1980 OECD privacy guidelines are available online at <http://www.oecd.org/document/18/0,2340,en_2649_201185_1815186_1_1_1_1,00.html>, accessed December 14, 2004.

[13] These five minimum guidelines are taken from "RFID Position Statement of Consumer Privacy and Civil Liberties Organizations," November 20, 2003, available online at <http://www.privacyrights.org/ar/RFIDposition.htm>, accessed December 14, 2004.

[14] See Ari Juels, Ronald L. Rivest, and Michael Szydlo, October 2003, "The Blocker Tag: Selective Blocking of RFID Tags for Consumer Privacy," pp. 103-111 in *Proceedings of the 10th ACM Conference on Computer and Communication Security*, ACM Press, New York.

for individuals as well as companies *not* to use the technology. Consider one of the primary applications of RFID currently in use: E-ZPass or FasTrak toll road systems in the eastern United States and California, respectively, allow cars to speed through tollbooths. Such applications, along with similar bus passes or gas-purchasing cards, have raised little skepticism. Grocery store loyalty cards that offer discounts, airport passes that divert "trusted travelers" through convenient screening, medical tags that ensure prescription tracking—such benefits will be difficult to refuse. These applications might involve a version of informed consent, but they can also result in unanticipated side effects. And the use of the tags without some of the measures described above means that subsequent tracking and profiling would be possible. Robust systems based in part on RFID will create an infrastructure that could serve many purposes, from government surveillance to targeted marketing by advertisers.

ESTABLISHING PUBLIC TRUST

Workshop panelists noted that what is most apparent in public discourse about emerging information technology is a lack of public trust. RFID technologies and systems will be most smoothly adopted and implemented if those involved take this issue seriously. Grappling with social and cultural considerations may alter the goals of scientists as well as manufacturers and retailers, but investment up front will make it more likely that significant dissonance can be avoided at a later stage (as in other research areas resulting in controversy such as agricultural biogenetics and stem cells). A key question is how to establish public trust as RFID technologies evolve and are implemented. As sociologist Anthony Giddens argues, contemporary society is riddled with abstract systems that require a nearly blind trust, akin to faith.[15] In order for emerging technology to earn this trust, workshop participants argued that safeguards, such as those mentioned above from the Fair Information Practice Principles, should be in place. To evolve RFID technology such that societal, commercial, and governmental interests work together will require significant and meaningful participation by a range of stakeholders, including advocates of the public interest.

A technological fix that begins to address primary concerns might be simply to kill all RFID tags at the point of sale and to make this operation straightforward and obvious. While some argue that this answer would eliminate many privacy threats that RFID poses, including the most chilling effects such as profiling and tracking, it raises other issues.[16] And what of consumers interested in doing their own tracking of products for reasons related to public interest issues, such as monitoring food spoilage or finding the source of defective products? Or consumers personally interested in using RFID tags and readers to identify spoiled medications, for example? In general, understanding the topology of the application space and how values (such as privacy) are contextualized will be important.[17] Some application areas could be characterized as having large benefits and small risks, and others the opposite. Understanding how to describe various application areas and where the boundaries between them are will be helpful moving forward.

Ethical, legal, and social issues related to RFID technologies reflect some of the same concerns raised about other emerging, wireless and embedded technologies. Public concern centers on risks to privacy, and a number of public interest groups have issued guidelines for

[15] Anthony Giddens. 1990. *The Consequences of Modernity*, Stanford University Press, Stanford, Calif.

[16] However, privacy advocacy groups cite examples of intrusive in-store surveillance, which makes the strategy of killing tags at the point of sale only a partial solution.

[17] Characteristics of application area requirements to consider might include these: mobility versus fixed tag location, the sort of data that are involved (ID only or not), whether the tags are associated with individuals, the ability to turn a tag off, the visibility of tags and readers, and so on.

policy and technology development. By their doing so, the notion of "privacy" has become a more complex and nuanced issue and, arguably, no longer the proper name for all of the societal concerns subsumed under it. No matter which list of concerns over RFID technology is chosen, themes that arise repeatedly in the literature include privacy, trust, safety, security, fairness, accountability, accessibility, reliability, and informed consent.[18] Another notion that has received little consideration is "publicity," the flip side of privacy—the notion that emerging technologies, including RFID, be developed consistent with an obligation to contribute to the shared, public sphere. Incorporating publicity is another means, like that of widening stakeholder participation, to enhance public acceptance of and trust in systems.

In the discussion of social norms, privacy advocates seek to have protections built in so that consumers can control the exposure of their identities. But it is obvious that there are no absolute norms that can be applied in contemporary society, with its great diversity. Consider individuals willing to have an RFID chip implanted for purposes such as clubbing (a rice-sized chip embedded in the upper arm allows its wearer to jump entry queues, reserve a table, or pay for drinks).[19] While some privacy advocates suggest that this type of use is undesirable over the long term, there are undoubtedly numerous closed-system contexts in which some people will wish to be "tagged." In some cases, trust may be the result of a negotiation between the provider and the user of the technology. A technology or service provider's reputation (regarding privacy, security, trustworthiness, and so on.) may become an important component of such negotiations.

Given the vast differences in individual preferences regarding privacy, along with a range of social norms, the establishment of public trust with respect to RFID technology will be a complicated, long-term undertaking. Indeed, it may be that trusted technology developers will hold a special corner on the market.[20] If RFID systems are not designed, developed, and deployed with public trust in mind, privacy advocates may feel the need to resort to less restrained efforts—worst-case scenarios hold powerful sway in the public imagination. Moreover, because privacy for individuals is the most well-articulated societal implication of RFID, the technology may be skewed in this direction. That means that the collective benefits that RFID systems might enable—for instance, bringing down prices on consumer goods, improving security in response to terrorist threats, and enhancing health and education applications—could be secondary to individual privacy goals. Thus, it was argued at the workshop, the desirability of developing RFID systems with societal concerns in mind is clear, and developing means to do so will be an important strategy for all stakeholders as the technology moves forward.

---

[18] One issue mentioned at the workshop does not seem to come up very frequently—the notion of environmental sustainability. A proposed 96-bit identity space would (conceptually) allow every person on the planet to have billions of billions of tags. Even though each tag is very small, numbers like this raise questions about reusability, reprogrammability, and recycling.

[19] Duncan Graham-Rowe, 2004, "Clubbers Choose Chip Implants to Jump Queues," *New Scientist*, May 21, available online at <http://www.newscientist.com/news/print.jsp?id=ns99995022>. See also Sherrie Gossett, 2004, "Paying for Drinks with Wave of the Hand," *WorldNet Daily*, April 14, available online at <http://worldnetdaily.com/news/article.asp?ARTICLE_ID=38038>, accessed December 14, 2004.

[20] One suggested possibility is to start assigning trust ratings, like a Good Housekeeping Seal or an e-Bay "feedback score," to RFID manufacturers, with high marks going to those that anonymize their data, demonstrate visibly that a tag is on, off, or killed, and so on.

Appendixes

A

Workshop Agenda

RADIO FREQUENCY IDENTIFICATION (RFID) TECHNOLOGIES: A WORKSHOP

May 10-11, 2004
Watertown Hotel
Seattle, Washington

Monday, May 10

8:45–9:00 a.m.

Introduction and Overview
Lynette Millett, Study Director, Computer Science and Telecommunications Board
Gaetano Borriello, Chair, Committee on Radio Frequency Identification Technologies

9:00–10:15

Session 1: Brief History and Overview of RFID Technology—Where We Stand
Moderator: Bill Schilit (Scribe: Gaetano Borriello)

- What are the technical realities of RFID? (What is its current functionality? What is it useful/not useful for?)
- What are adjacent technologies and their complementary/competing roles (e.g., contactless smart cards and active tags)?
- What is the spectrum of RFID and RFID-related technologies?

Panelists: Kevin Ashton, Tim Harrington, Roy Want

10:15–10:30

Break

10:30–12:30 p.m.

Session 2: Business Case for and Against RFID Technologies
Moderator: Paul Zipkin (Scribe: Bill Schilit)

- What are the business implications of RFID technologies? (For example, how does RFID work in the supply chain? What are the economics of it? How is it likely to be used in industry?)

- What are the possible implications of mandates from government and business?
- What are potential or actual consumer RFID products?
- What are the important price points for RFID technology, including tags and readers?
- What are other economic factors that need to be taken into account?

Panelists: Bruce Eckfeldt, Ted Klastorin, Eric Peters, Ravi Rajapakse, Sandy Williamson

12:30–1:00	Lunch

1:00–3:00 Session 3: Where the Technology Is Going
Moderator: Steven Shafer (Scribe: Chris Diorio)

- What are some of the developing technical aspects of RFID technology (e.g., blocker tags, enhanced antenna design, positioning technologies, encryption and other data safeguards, etc.)?
- What are some of the near-term (i.e., within 5 years) expectations for RFID technology?
- What factors are limiting or promoting research and development?

Panelists: Dan Bailey, Tim Harrington, Ravi Pappu, Ravi Rajapakse, Louise Sengupta

3:00–3:30 Break

3:30–5:00 Session 4: RFID Infrastructure and Data Management Issues
Moderator: Gaetano Borriello (Scribe: Paul Zipkin)

- How might database technology evolve to handle all the data from RFID transactions?
- What are the ramifications for data distribution, networking, storage, mining, and so on?
- What are the systems and infrastructure issues that need to be addressed to make these challenges manageable?

Panelists: Greg Pottie, Sumit Roy, Javed Sikander, Jim Waldo

Tuesday, May 11th

8:30–10:30 a.m. Session 5: Privacy, Social, and Cultural Concerns
Moderator: Dana Cuff (Scribe: Steve Shafer)

- How might the use and distribution of RFID technology affect personal privacy and anonymity?
- How will the interests and rights of consumers be handled?
- What are possible government and law enforcement uses of RFID transaction data?
- What effects will worldwide standards and other bodies have on RFID privacy issues?
- What are the potential social and cultural implications of significant RFID use?
- What are the data policy implications of RFID technologies (e.g., control, access, and ownership of RFID-generated data, etc.)?

	Panelists: Paula Bruening, Kenneth Fishkin, Batya Friedman, Ravi Pappu, Lee Tien
10:30–11:00	Break
11:00–12:30 p.m.	Session 6: RFID, Government, and Standards Moderator: Chris Diorio (Scribe: Dana Cuff) • What are the laws, standards, and regulations (if any) surrounding RFID technology, and its development and use? • How do spectrum policy considerations both in the United States and abroad relate to RFID technology? • What government agencies have oversight responsibilities that could affect the development and deployment of RFID technologies? Panelists: Kevin Ashton, Harley Heinrich, Lauren Van Wazer
12:30–1:00	Lunch
1:00–2:15	Session 7: Looking to the Future, Part 1—Predictive Moderator: Gaetano Borriello (Scribe: Staff) • What are likely useful consumer applications of RFID technologies? • What are likely business and government applications? • What are some anticipated technical and social challenges? Panelists: Bruce Eckfeldt, Jim Waldo, Roy Want
2:15–3:30	Session 8: Looking to the Future, Part 2—Speculative Moderator: Gaetano Borriello (Scribe: Staff) • What are some speculative applications for RFID technologies that have not been discussed much at this or other similar workshops or in the literature? Brainstorming session with all participants.
3:30	Adjourn

B

Biosketches of Committee Members

GAETANO BORRIELLO is a professor in the Department of Computer Science and Engineering at the University of Washington. He has a B.S. in electrical engineering from the Polytechnic Institute of New York (1979), an M.S. in electrical engineering from Stanford University (1981), and a Ph.D. in computer science from the University of California, Berkeley (1988). He was a member of the research staff at the Xerox Palo Alto Research Center from 1980 to 1987. Dr. Borriello is known primarily for his work in automatic synthesis of digital circuits, reconfigurable hardware, and embedded systems development tools. He recently was principal investigator for the Portolano Expedition, an investigation on invisible computing that was sponsored by the Defense Advanced Research Projects Agency (DARPA). Dr. Borriello was on partial leave from 2001 to 2003 to found and direct the Intel Research Seattle laboratory, which is engaged in ubiquitous computing research. His research interests focus on location-based systems, sensor-based inferencing, and tagging objects with passive and active tags. Dr. Borriello has served as program chair of numerous conferences and workshops. His most recent community activities include being program chair for the 4th International Conference on Ubiquitous Computing, serving on the editorial board of *IEEE Pervasive Computing* magazine, and contributing to the National Research Council study that produced *Embedded, Everywhere: A Research Agenda for Networked Systems of Embedded Computers* (National Academy Press, 2001).

DANA CUFF is a professor in the Department of Architecture and Urban Design at the University of California, Los Angeles. She received her B.A. from the University of California, Santa Cruz, and her Ph.D. from the University of California, Berkeley. Dr. Cuff is a founder and director of the Institute for Pervasive Computing and Society (iPerCS), and a member of the Center for Embedded Networked Sensing faculty at UCLA. Dr. Cuff's work focuses on the cultural production of architecture and the city. She has published and lectured widely on emerging technology's impact on architecture and urbanism, the architectural profession, affordable housing, and the politics of place. Her most recent book, *The Provisional City: Los Angeles Stories of Architecture and Urbanism* (The MIT Press, 2001) was supported by both the J. Paul Getty Trust and the National Endowment for the Arts. She is currently researching ubiquitous computing technologies and their impact on the public sphere. An article entitled "Pervasive Computing: Embedding the Public Sphere," written with iPerCS cofounder Jerry Kang, was published in the *Washington and Lee Law Review* (Spring 2005). Dr. Cuff was awarded a Humanities Research Institute Fellowship for the year 2004 to examine the evolution of the neighbor and neighborhood in postwar American suburbs, including the growing impacts of pervasive computing on everyday life.

CHRIS DIORIO is an associate professor of computer science and engineering at the University of Washington and is a cofounder of Impinj, Inc., Seattle, Washington. Dr. Diorio is the co-chair of the EPCglobal Hardware Action Group and has worked actively in the development of RFID standards. He has received several awards, including the University of Washington Distinguished Teaching Award in 2001, an ONR Young Investigator Award in 2001, an Alfred P. Sloan Foundation Research Fellowship in 2000, a Presidential Early Career Award in Science and Engineering (PECASE) in 1999, a Packard Foundation Fellowship in 1998, and the Electron Devices Society's Paul Rappaport Award in 1996. He has worked as a senior staff engineer at TRW, Inc., as a senior staff scientist at American Systems Corp., and as a technical consultant at The Analytic Sciences Corp. He received his B.A. in physics from Occidental College in 1983 and his M.S. and Ph.D. in electrical engineering from California Institute of Technology (Caltech) in 1984 and 1997, respectively.

BILL SCHILIT is co-director of Intel Corporation's Intel Research Seattle and is part of a small team chartered with defining and driving Intel's ubiquitous computing agenda. Dr. Schilit's research focuses on ubiquitous and proactive computing applications, with an emphasis on context-aware computing. His research is positioned at the intersection of networking and human-computer interaction. Prior to joining Intel, he managed the Personal and Mobile Computing Group at FX Palo Alto Laboratory, a Fuji Xerox company. Dr. Schilit also worked at AT&T Bell Laboratories and Xerox Palo Alto Research Center (PARC). At PARC, he championed the notion of location-aware computing, coined the term "context-aware computing," and helped invent, design, and build the software and applications for the PARCTAB. He is associate editor-in-Chief of *IEEE Computer*, an area editor of *IEEE Wireless Communications*, and a member of the IEEE Computer Society and the Association for Computing Machinery.

STEVEN SHAFER is a senior researcher at Microsoft Corporation, working in the area of ubiquitous computing. He received his B.A. from the University of Florida in 1976 and his Ph.D. from Carnegie Mellon University (CMU) in 1983. He was then a faculty member at Carnegie Mellon until 1995. Dr. Shafer founded the Calibrated Imaging Laboratory, working on the modeling of color, highlights, texture, and lens and camera optics. He also worked on robot driving in the CMU Navigation Laboratory robot truck project. Dr. Shafer was a founder and later chair of the robotics doctoral program at Carnegie Mellon, and he helped establish the Human-Computer Interaction Institute. Dr. Shafer joined Microsoft in 1995, where he started the EasyLiving project to develop an architecture for building intelligent environments. His current work is in location awareness and radio frequency identification (RFID). He is past chair of the IEEE Pattern Analysis and Machine Intelligence Technical Committee, the primary scientific organization for computer vision, and an associate editor for the *IEEE Pervasive Computing* magazine, and he serves on the program committees of numerous recent conferences in pervasive and ubiquitous computing. Dr. Shafer is one of the Microsoft representatives at EPCglobal, an international RFID standards organization.

PAUL ZIPKIN is the T. Austin Finch, Sr., Professor at the Fuqua School of Business, Duke University. He received his Ph.D. from Yale University in 1977. His teaching, research, and consulting focus on how supply chains work and on how to make them work better, as well as on their strategic roles in the success or failure of companies in the global marketplace. Within this broad theme, Dr. Zipkin's work is concerned with issues of inventory management in supplier-customer relations; the impact of new production and communications technologies on supply-chain performance; coping with product variety at both the operational and strategic levels; and the design of logistics networks. He has published some 50 articles in scholarly journals and co-edited the book *Logistics of Production and Inventory* (Elsevier, 1993). He is the author of the book *Foundations of Inventory Management* (McGraw-Hill, 2000). Dr. Zipkin often advises companies, government agencies, and other organizations.

What Is CSTB?

As a part of the National Research Council, the Computer Science and Telecommunications Board (CSTB) was established in 1986 to provide independent advice to the federal government on technical and public policy issues relating to computing and communications. Composed of leaders from industry and academia, CSTB conducts studies of critical national issues and makes recommendations to government, industry, and academia. CSTB also provides a neutral meeting ground for consideration of complex issues where resolution and action may be premature. It convenes discussions that bring together principals from the public and private sectors, assuring consideration of key perspectives. The majority of CSTB's work is requested by federal agencies and Congress, consistent with its National Academies context.

A pioneer in framing and analyzing Internet policy issues, CSTB is unique in its comprehensive scope and its effective, interdisciplinary appraisal of technical, economic, social, and policy issues. Beginning with early work in computer and communications security, cyber-assurance and information systems trustworthiness have been a cross-cutting theme in CSTB's work. CSTB has produced several reports known as classics in the field, and it continues to address these topics as they grow in importance.

To do its work, CSTB draws on some of the best minds in the country and from around the world, inviting experts to participate in its projects as a public service. Studies are conducted by balanced committees without direct financial interests in the topics they are addressing. Those committees meet, confer electronically, and build analyses through their deliberations. Additional expertise is tapped in a rigorous process of review and critique, further enhancing the quality of CSTB reports. By engaging groups of principals, CSTB gets the facts and insights critical to assessing key issues.

The mission of CSTB is to

- *Respond to requests* from the government, nonprofit organizations, and private industry for advice on computer and telecommunications issues and from the government for advice on computer and telecommunications systems planning, utilization, and modernization;
- *Monitor and promote the health of the fields* of computer science and telecommunications, with attention to issues of human resources, information infrastructure, and societal impacts;
- *Initiate and conduct studies* involving computer science, technology, and telecommunications as critical resources; and
- *Foster interaction* among the disciplines underlying computing and telecommunications technologies and other fields, at large and within the National Academies.

CSTB projects address a diverse range of topics affected by the evolution of information technology. Recently completed reports include *Getting Up to Speed: The Future of Supercomputing; Cybersecurity Today and Tomorrow: Pay Now or Pay Later*; *Computer Science: Reflections on the Field, Reflections from the Field; Youth, Pornography, and the Internet*; *IDs—Not That Easy: Questions About Nationwide Identity Systems*; *The Internet Under Crisis Conditions: Learning from September 11*; and *Innovation in Information Technology.* For further information about CSTB reports and active projects, see <http://cstb.org>.